AQA GCSE (9–1) Chemistry

Required Practicals Lab Book

Emily Quinn

William Collins' dream of knowledge for all began with the publication of his first book in 1819. A self-educated mill worker, he not only enriched millions of lives, but also founded a flourishing publishing house. Today, staying true to this spirit, Collins books are packed with inspiration, innovation and practical expertise. They place you at the centre of a world of possibility and give you exactly what you need to explore it.

Collins. Freedom to teach

Published by Collins
An imprint of HarperCollins*Publishers*
The News Building
1 London Bridge Street
London SE1 9GF

Browse the complete Collins catalogue at
www.collins.co.uk

15 14 13

ISBN 978-0-00-829162-4

British Library Cataloguing in Publication Data
A catalogue record for this publication is available from the British Library.

Author: Emily Quinn
Commissioning editor: Rachael Harrison
In-house project editor: Isabelle Sinclair
Copyeditor and typesetter: Hugh Hillyard-Parker
Proofreader and answer checker: Stuart Lloyd
Artwork: QBS Learning
Cover designer: Julie Martin
Cover photo: © Sebastian Janicki/Shutterstock.com
Production controller: Tina Paul
Printed and bound by: Martins the Printers, Berwick-upon-Tweed

MIX
Paper from responsible sources
FSC™ C007454

This book is produced from independently certified FSC™ Paper to ensure responsible forest management.

For more information visit: www.harpercollins.co.uk/green

The publishers will gladly receive any information enabling them to rectify any error or omission at the first opportunity.

Contents

How to use this book

Practical skills are at the heart of any science qualification. Your AQA GCSE Science course requires you to develop these skills through completing a series of required practicals, which you will then be tested on in your exams. This lab book will help you record the results of your practical work, and provide you with some guidance so that you get the most out of your time completing each practical.

Ensure you write down everything you can about your practical work – remember you can refer back to this book when you're revising!

Learning outcomes

This is a summary of what you should have accomplished by the end of each required practical.

Apparatus list

Your teacher will ensure that all the apparatus you need for the practical can be found in the classroom. You can use this list to check that you have everything you need to start your work.

Maths skills required

This is a good reminder of the skills you will need to master and practise on your science course, which will be tested in your exams. There are also questions included throughout that let you practise your maths skills.

Formulae

Any formulae you need to know to complete your practical work are shown here.

Safety notes

You should always be aware of safety when completing any practical work. This list will help you be aware of any common safety issues! Your teacher will advise on safety information for each practical, so pay attention.

Common mistakes

We've included some of the common mistakes people make during their practical work so that you can look out for them and hopefully avoid making the same mistake!

Method

Always make sure you read every step of the method before you begin work. This will help you avoid mistakes and will give you an idea of what outcomes to look for as you complete each step.

Record your results

At the end of every method is place to record the outcomes of your work. Make sure you keep your notes clear and neat.

Check your understanding and **Exam-style questions**

For each practical, there are questions designed to check your understanding of the work you've just completed. There are also exam-style questions included to help you prepare for questions in the exams. Some of these questions are designed to test your maths skills and to check your understanding of the apparatus and techniques that you've been using – you'll be tested on these in your exams.

Higher Tier

HT If you see this symbol next to a question, then it is designed for Higher Tier content only.

Teachers should always ensure they consult the latest CLEAPSS safety guidance before undertaking any practical work.

4.4.2.3 Making salts

Being able to choose the correct techniques and carry out a specified procedure to produce a pure product is an important skill for scientists. You will probably know how to complete each of the necessary techniques separately, but can you explain how to use them together to produce a product safely?

Your task is to prepare a pure, dry sample of a soluble salt from an insoluble oxide or carbonate, using a Bunsen burner to heat dilute acid and a water bath or electric heater to evaporate the solution.

You will react an acid and an insoluble oxide or carbonate to prepare an aqueous solution of a salt. The unreacted base from the reaction will need to be filtered. You will evaporate the filtrate to leave a concentrated solution of the salt, which will crystallise as it cools and evaporates further. When dry, the crystals will have a high purity.

Learning outcomes	Maths skills required	Formulae
• React a metal oxide with an acid to make a salt. • Work safely with hot chemicals. • Write word and chemical equations for this reaction.	• Measure precisely using a measuring cylinder. • Discuss resolution of practical equipment.	• M_r of a compound = A_r of all the elements present in the compound

Apparatus list

- 40 cm^3 1.0 M dilute sulfuric acid (M = mol/dm^3)
- 5 g copper (II) oxide powder
- spatula
- glass rod
- 100 cm^3 beaker
- conical flask
- Bunsen burner
- tripod
- gauze
- heatproof mat
- filter funnel
- filter paper
- 250 cm^3 beaker
- evaporating basin
- crystallising dish
- tongs
- eye protection

If crystallising dishes are not available, Petri dishes (without lids) make good substitutes.
If small conical flasks are not available, a second small beaker is an acceptable replacement.

Safety notes

- Wear eye protection at all times.
- **Do not** boil the acid! It can release harmful sulfur dioxide gas!
- Don't add large amounts of copper oxide in one go – the solution will boil and bubble over.
- If you allow the water bath to boil dry, it may cause the beaker to crack.
- You should not take the crystals home or pour copper sulfate down the sink.
- When lifting your hot beaker with tongs, ensure you use these on the rim of the beaker to avoid slipping.

Always be careful when handling chemicals and follow your teacher's safety advice about the below:

- 40 cm^3 1.0 M dilute sulfuric acid *(irritant)*
- copper (II) oxide powder *(harmful)*
- copper (II) sulfate *(corrosive)*

Common mistakes

- You might struggle to heat the solution gently. Ask your teacher for a demonstration of how to heat using a gentle flame – with the air hole partly closed and the gas tap half open.
- If the solution starts to spit, move your Bunsen burner from underneath the solution.
- You might struggle to see where the crystallisation point is in step **6** below. You are looking for tiny crystals forming on the glass rod.
- In step **4** of the method, the filtered solution should be clear. If there is any black copper oxide in it then you should filter the solution again.

Method

Read these instructions carefully before you start work.

1. Pour 40 cm^3 of 1.0 M sulfuric acid into a 100 cm^3 beaker.
2. Set up your Bunsen burner on a heatproof mat with a tripod and gauze over it. Place the beaker with the acid on the gauze and heat gently over a gentle flame until almost boiling. **Turn the Bunsen burner off**.
3. Carefully place the hot beaker on the heatproof mat using tongs. Add the copper oxide one spatula at a time, stirring using the glass rod as you add it. The mixture will turn clear and blue.
4. Fold and place a filter paper inside a funnel and filter the blue copper sulfate solution into a conical flask.

Figure 1

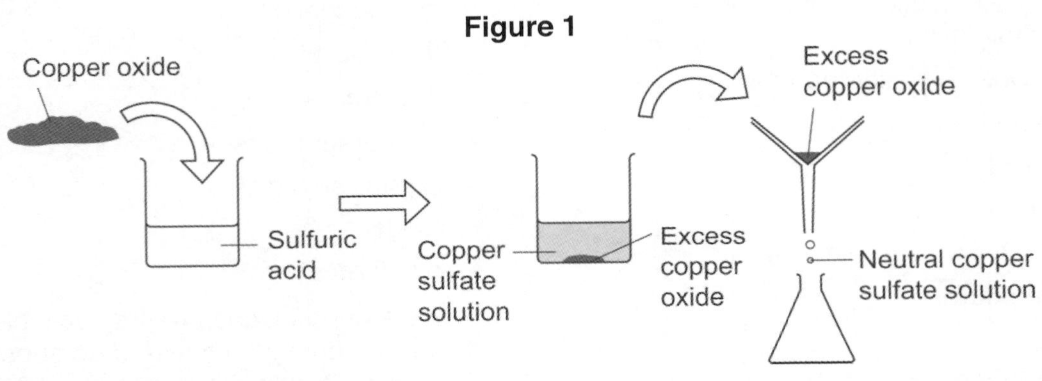

5. Pour the copper sulfate solution into the evaporating basin and heat over a 250 cm^3 beaker acting as a water bath until about half the solution has evaporated.
6. Test the solution by dipping a clean glass rod into it and then letting the rod cool. When small crystals form on the glass rod, stop heating the solution.
7. Use tongs to pour the copper sulfate solution into a crystallising dish and leave it in a warm place to finish crystallising.

Record your results

Write the word equation for this reaction.

....

Write the chemical equation for this reaction.

....

Describe the properties of the reactants and products, i.e. colour, state etc., in **Tables 1** and **2** below.

Table 1 – Descriptions of reactants

Reactant	Chemical formula	Properties
Sulfuric acid		
Copper (II) oxide		

Table 2 – Descriptions of products

Product	Chemical formula	Properties
Copper sulfate crystals		
Water		

Check your understanding

1. Copper oxide is used in excess in the reaction with sulfuric acid.

a. State what 'in excess' means. [1 mark]

....

....

Copper oxide is used in excess to ensure this practical is safe.

b. Describe what could happen if copper oxide was not used in excess. [1 mark]

....

2. Copper oxide (CuO) is reacted with sulfuric acid (H_2SO_4) to make copper sulfate ($CuSO_4$) and water (H_2O).

a. Calculate the total relative formula mass (M_r) for the reactants of this reaction. [3 marks]

....

....

....

b. State the law of conservation of mass. [1 mark]

....

c. Explain how the law of the conservation of mass allows you to predict the total M_r of the products.

Use this law to predict the M_r of the products. [2 marks]

....

....

Exam-style questions

1. Copper oxide reacts with hydrochloric acid to form crystals of a new substance.

   a. State the name of the new substance formed. [1 mark]

   ....

   b. Suggest a method for obtaining a pure sample of crystals of this new substance. [6 marks]

   ....

   ....

   ....

   ....

   ....

   ....

2. A student reacts calcium carbonate with hydrochloric acid.

   a. Complete the symbol equation for this reaction: [1 mark]

   $CaCO_3 + 2HCl \rightarrow CaCl_2 + H_2O +$ ....

   b. Name the salt made in this reaction. [1 mark]

   ....

3. Another way to make a salt is by adding a metal to an acid.

   a. Suggest the reactants needed to make the salt sodium nitrate. [1 mark]

   ....

   ....

   b. Suggest why this reaction would not be safe to do in the lab. [1 mark]

   ....

   ....

4.4.2.4 Neutralisation

Titration is a useful analytical tool to find the concentration of unknown substances. Indicators can be used to show when an acid and alkali have neutralised each other. Neutralisation reactions produce salt and water that have a pH of 7. In this practical you are going to determine the volumes of reacting solutions of a strong acid and a strong alkali by titration.

HT

You will also determine the concentration of one of the solutions in mol/dm^3 and g/dm^3 from the reacting volumes and the known concentration of the other solution.

Learning outcomes	Maths skills required	Formulae
• Carry out a titration safely and accurately. • Use the burettes correctly. • **HT** Calculate the concentration of a solution from the concentration and volume of another solution.	• **HT** Calculate moles of one solution by using the concentration and volume of the other and a balanced symbol equation. • Calculate a mean.	• **HT** $\text{Concentration (mol/dm}^3\text{)} = \dfrac{\text{number of moles}}{\text{volume of solution (dm}^3\text{)}}$

Apparatus list

- 25 cm^3 volumetric pipette and pipette filler
- burette
- clamp stand
- 250 cm^3 conical flask
- white tile or white paper
- 0.4 M hydrochloric acid
- 0.4 M sodium hydroxide solution (**HT** of unknown concentration)
- methyl orange indicator
- eye protection

Safety notes

- Wear eye protection at all times!
- Be careful with the burettes, especially the ends. They snap off easily and are expensive.

Always be careful when handling chemicals and follow your teacher's safety advice about the below:

- 0.4 M hydrochloric acid *(low hazard)*
- 0.4 M sodium hydroxide *(irritant)*
- **HT** unknown concentration sodium hydroxide *(treat as if corrosive)*
- Methyl orange indicator *(toxic)*

Common mistakes

- Put the ACID in the burette, not the alkali. Alkali is very hard to clean out of the burettes.
- Practise adding acid dropwise (drop by drop) to make sure you have an accurate measurement for the amount of acid used.
- Practise swirling with one hand and using the burette stopper with the other OR work in pairs with one person swirling and the other controlling the stopper.
- The alkali has been neutralised when the methyl orange changes from yellow orange to red completely. A colour change that disappears on swirling means that the solution is not yet neutral.
- Make sure you measure the volume in the burette from the bottom of the meniscus.

Method

Read these instructions carefully before you start work.

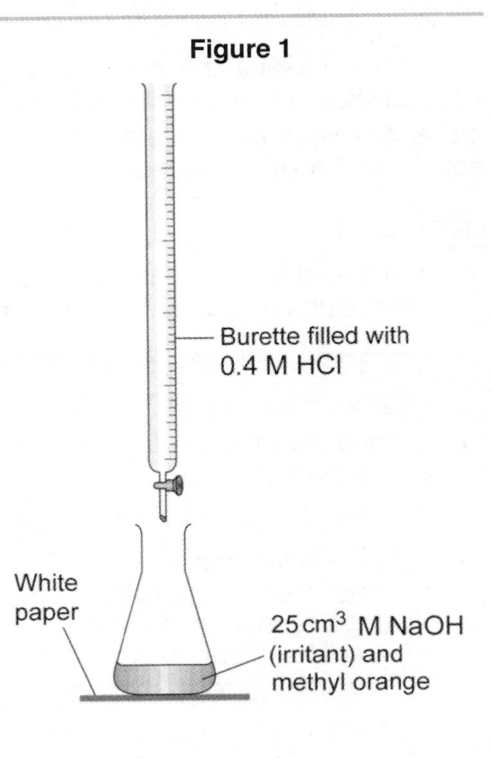

1. Set up the equipment as shown in **Figure 1** with a clamp stand holding the burette.
2. First find the approximate volume of acid needed to neutralise the alkali. To do this, pour 25 cm^3 of sodium hydroxide, NaOH, into a conical flask. Run the hydrochloric acid, HCl, into the sodium hydroxide slowly, always swirling, until the solution just turns from yellow-orange to red. Record the result in the 'approximate' column in **Table 1**.
3. Empty the flask and rinse it with distilled water.
4. For **Trial 1** measure out 25 cm^3 of sodium hydroxide, NaOH, and repeat the titration. Add the hydrochloric acid one drop at a time near the expected end-point.
5. For **Trial 2** and **Trial 3**, repeat steps **2** and **3** until consistent results are obtained.
6. Calculate the mean volume of 0.4 M hydrochloric acid, HCl, needed to neutralise the alkali.

Record your results

Table 1 – Neutralisation results

Burette reading (cm^3)	Approximate volume (cm^3)	Trial 1 (cm^3)	Trial 2 (cm^3)	Trial 3 (cm^3)
Initial				
Final				
Titre (amount used)				

Calculating concentrations for unknowns (HT only)

Step 1

$$\text{Concentration (mol/dm}^3) = \frac{\text{number of moles}}{\text{volume of solution (dm}^3)}$$

Calculate the number of moles of sodium hydroxide in 25 cm^3 of (for example) 0.4M NaOH.
(Remember to convert cm^3 to dm^3 by dividing by 1000.)

Step 2

If required, balance the equation for the reaction:

.... NaOH + ... HCl → .... NaCl + ... H_2O

Step 3

Identify how many moles of acid are needed to neutralise the number of moles of alkali (for example, a 1:1 ratio). You do this by looking at the stoichiometry of the balanced equation:

Moles of acid = ...

Moles of alkali = ...

Step 4

$$\text{Concentration (mol/dm}^3\text{)} = \frac{\text{number of moles}}{\text{volume of solution (dm}^3\text{)}}$$

Use the number of moles of acid and the **mean volume used** to calculate the concentration of the acid.

Step 5

Calculate the amount of the acid in grams.

$$\text{Number of moles} = \frac{\text{mass of substance (g)}}{M_r \text{ of substance}}$$

Calculate the M_r of the acid and use the number of moles to find the mass.

Check your understanding

1. Experiments like this require you to be accurate and precise.

   a. Define the word 'accurate'. [1 mark]

   ....

   b. Define the word 'precise'. [1 mark]

   ....

   c. Describe how precise your results are. [1 mark]

   ....

d. Explain why the use of burettes in this experiment helps you to gain precise results. [1 mark]

....

Exam-style questions

1. A student performs a titration between sulfuric acid (H_2SO_4) and 25 cm^3 of 0.5 M sodium hydroxide (NaOH).

Table 2 shows their results.

Table 2

Burette reading (cm^3)	**Approximate volume (cm^3)**	**First trial (cm^3)**	**Second trial (cm^3)**	**Third trial (cm^3)**
Initial	1.5	1.8	2.4	15.2
Final	28.5	28.4	29.1	41.8
Titre (amount used)	27.0	26.6	26.7	

a. Calculate the titre (amount used) in the third trial. [1 mark]

....

b. Calculate the mean trial titre. [1 mark]

....

c. Calculate the uncertainty of the mean titre. [1 mark]

....

2. **HT** Calculate the concentration of the sulfuric acid used to neutralise the sodium hydroxide.

Use the formula:

$$\text{Concentration (mol/dm}^3\text{)} = \frac{\text{number of moles}}{\text{volume of solution (dm}^3\text{)}}$$

[3 marks]

....

....

....

4.4.3.4 Electrolysis

Electrolysis uses electricity ('electro-') to split ('-lysis') compounds in two. You are going to investigate what happens when aqueous solutions are electrolysed using inert electrodes and find out what products are made. You will develop a hypothesis using your knowledge of reduction and oxidation and the reactivity series.

All the solutions contain H^+ ions and OH^- ions from partially ionised water molecules as well as the metal ions and non-metal ions from the soluble salt. For each electrolyte (the substance you perform electrolysis on) you will predict which element will be produced at the cathode and the anode, and then see if your observations support your predictions.

Learning outcomes	Maths skills required
• Devise a hypothesis. • Safely carry out electrolysis. • Identify elements produced from your observation. • **HT** Apply your knowledge of oxidation and reduction.	• **HT** Balance half equations.

Apparatus list

- copper (II) chloride solution
- copper (II) sulfate solution
- sodium chloride solution
- sodium sulfate solution
- 100 cm^3 beaker
- Petri dish lid
- eye protection
- two carbon rod electrodes
- two crocodile clips
- two leads
- 6 V power supply
- blue litmus paper
- tweezers

Safety notes

- Wear eye protection at all times!
- Chlorine is produced during two of the experiments. **Do not inhale it!** Chlorine gas (Cl_2) reacts with water (H_2O) on the mucous membranes in the lungs and produces hydrochloric acid (HCl), which then damages tissue in the lungs.
- Make sure the laboratory is well ventilated (open the windows and put the extraction fan on).
- **Do not** have a potential difference of higher than 4 V – too much chlorine will be made.
- **Do not** run the electrolysis for more than 5 minutes – too much chlorine will be made.
- If any of the equipment is damaged, **do not** use it.

Always be careful when handling chemicals and follow your teacher's safety advice about the below:

- copper (II) chloride solution *(harmful)*
- copper (II) sulfate solution *(harmful)*
- sodium chloride solution *(low hazard)*
- sodium sulfate solution *(low hazard)*

Common mistakes

- Do not let the electrodes touch as you will create a circuit and no products will be formed.
- Make sure you use a d.c. power supply.
- Make sure you know which electrode is connected to which terminal (positive or negative).

Method

Read these instructions carefully before you start work.

Figure 1

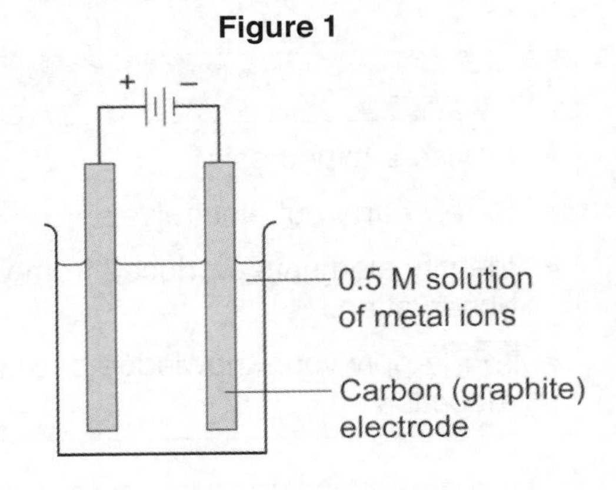

1. Make a hypothesis about what products will be formed at each electrode and write this in **Record your results**.
2. Measure 50 cm^3 of copper (II) chloride solution and pour into the beaker.
3. Place the lid with carbon rods over the beaker OR attach the carbon rods to the edge of the beaker using crocodile clips. **The rods must not touch each other**.
4. Attach leads to the rods/crocodile clips and connect the rods to the **direct current (red and black) terminals** of a power pack (see **Figure 1**).
5. Set the potential difference at 4 V on the power pack and switch on for **5 minutes maximum**. You may see bubbles appearing very quickly.

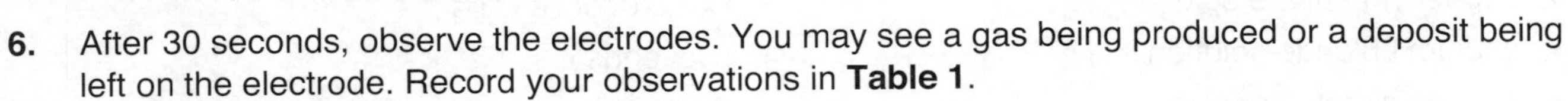

6. After 30 seconds, observe the electrodes. You may see a gas being produced or a deposit being left on the electrode. Record your observations in **Table 1**.

   If a gas is produced, test it by holding a piece of damp litmus paper in it with tweezers.

   If the paper is bleached, you have a positive identification of an element.
   Record the result in **Table 1**.
7. After five minutes, **switch off the power supply.**

   Examine the surface of the electrodes for a coating. If there is a coppery-red deposit, you have a positive identification of an element. Record your results in **Table 1**.
8. Clean the equipment making sure there are no deposits left on the electrodes.
   Rinse with **distilled** water, not tap water.
9. Repeat steps **1**–**8** using:
   - copper (II) sulfate
   - sodium chloride
   - sodium sulfate.

Note: Depending on the substance in the solution, other substances may be formed. Gas produced at the positive electrode that does **not** bleach blue litmus paper is oxygen. If you collect the oxygen and add a glowing splint to it, it will relight it. If a gas is produced at the negative electrode, it is hydrogen. If you collect the hydrogen and add a lit splint to it, it will make a pop.

Record your results

Hypotheses

Table 1 – Products of electrolysis

Solution	Positive electrode (anode)		Negative electrode (cathode)	
	Observations	Element formed	Observations	Element formed
Copper (II) chloride				
Copper (II) sulfate				
Sodium chloride				
Sodium sulfate				

Check your understanding

1. Copper (II) chloride solution can be electrolysed to produce a copper deposit and chlorine gas.

   At which electrode will copper collect? ....

   At which electrode will chlorine gas be produced? ....

   [1 mark]

2. **HT** During electrolysis, copper (II) chloride solution forms Cu^{2+} ions and Cl^{-} ions.

   a. What is the formula of copper chloride? [1 mark]

   ....

   b. How many electrons does Cu^{2+} gain at its electrode? [1 mark]

   ....

   c. Is Cu^{2+} reduced or oxidised? [1 mark]

   ....

3. During electrolysis of sodium chloride, two gases are collected.

   a. Describe how a student could identify both of these gases. [2 marks]

   ....

   ....

   b. Describe any precautions the student should take to minimise the risks during this experiment. [2 marks]

   ....

   ....

Exam-style questions

1. **Table 2** shows the results of electrolysis for a range of compounds.

Table 2

Solution	Positive electrode (anode)		Negative electrode (cathode)	
	Observations	**Element formed**	**Observations**	**Element formed**
Copper (II) chloride	gas produced	chlorine	red/brown metal deposit on electrode	copper
Lithium chloride	gas produced	chlorine	gas produced	hydrogen
Sodium chloride	gas produced		gas produced	

   a. Name the elements formed during the electrolysis of sodium chloride solution, indicating the electrode at which each of them forms. [2 marks]

   Write your answers in **Table 2**.

   b. Predict the products of the electrolysis of potassium iodide solution.

   Explain why you have made this prediction. [4 marks]

   ....

   ....

   ....

   ....

   c. Predict the name of the solution left in the water. [1 mark]

   ....

2. During the electrolysis of aqueous copper (II) chloride, copper metal and chlorine gas are formed.

 a. **HT** Describe the reaction that occurs at the negative electrode that allows copper metal to be formed.

 Include a half equation for the reaction. [2 marks]

 ....

 ....

The electrolysis of molten lead bromide can be carried out using the equipment in **Figure 2**.

Figure 2

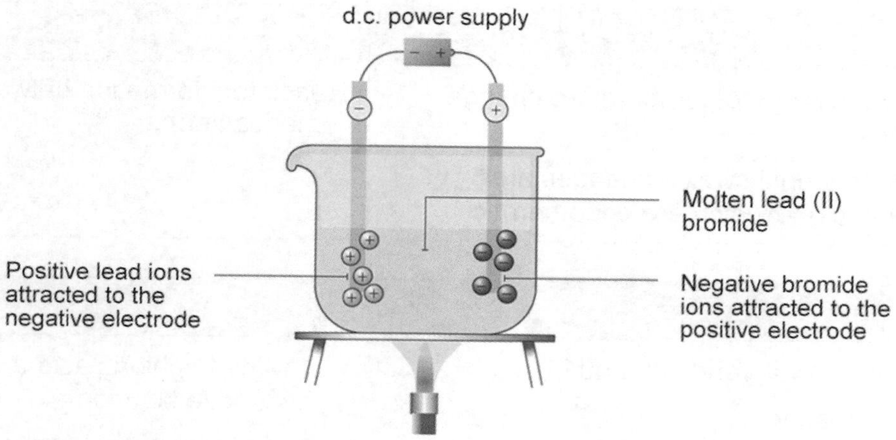

 b. Explain why the lead bromide must be molten. [2 marks]

 ....

 ....

 c. Lead is produced at the negative electrode when molten lead bromide is used, but hydrogen is produced when aqueous lead bromide is used.

 Explain why. [3 marks]

 ....

 ....

 ....

4.5.1.1 Temperature changes

Reactions can be endothermic (take in energy from the surroundings) or exothermic (give out energy to the surroundings). You will investigate a range of variables that affect temperature changes in reacting solutions such as:

1. neutralisation reactions
2. reactions between acids and carbonates
3. reactions between acids and metals
4. reactions involving displacement of metals.

You will monitor the temperature change as these reactions occur. The reaction will be contained in an insulated cup to reduce heat loss.

Learning outcomes	Maths skills required
• Carry out a range of temperature change reactions. • Be able to identify which changes are exothermic and which are endothermic.	Translate information between graphical and numerical form.

Apparatus list

- expanded polystyrene cup and lid
- 250 cm^3 beaker
- 10 cm^3 measuring cylinder
- 50 cm^3 measuring cylinder
- thermometer or temperature sensor
- eye protection
- spatula
- graph paper
- glass rod
- 1 M hydrochloric acid and 1 M sodium hydroxide solution
- 1 M ethanoic acid and a range of powdered metal carbonates (see CLEAPSS guidance)
- 0.5 M hydrochloric acid, strips of magnesium and pieces of zinc
- 1 M copper (II) sulfate solution and a range of metals

Safety notes

- Wear eye protection when doing all the practical activities and clearing up.
- Avoid skin contact with all the reactants.
- Some reactions can be **very** exothermic! **Do not touch** hot reactions and take care not to get burnt.

Always be careful when handling chemicals and follow your teacher's safety advice about the below:

- 1 M hydrochloric acid *(low hazard)*
- 1 M sodium hydroxide solution *(corrosive)*
- 1 M ethanoic acid (*low hazard*)
- 0.5 M hydrochloric acid (*low hazard)*
- 1 M copper (II) sulfate solution (*harmful)*
- powdered metal carbonates (see CLEAPSS guidance)

Common mistakes

- 30 cm thermometers are easier to read than 15 cm. In this experiment, shorter thermometers would have most of the scale below the lid of the polystyrene cup so would be harder to read.

- You can use other material such as wood or cardboard if you don't have polystyrene cup lids, as long as the cup is kept covered to reduce heat loss by convection.
- Be sure to keep the bulb end of the thermometer in the liquid while taking measurements.
- When the reaction takes place, stir the mixture to distribute the heat evenly.
- Ask your teacher if you need help drawing two lines of best fit so that they intersect.

Methods

Read these instructions carefully before you start work.

There are four activities to complete.

Start by setting up the equipment as shown in **Figure 1** or in **Figure 2**.

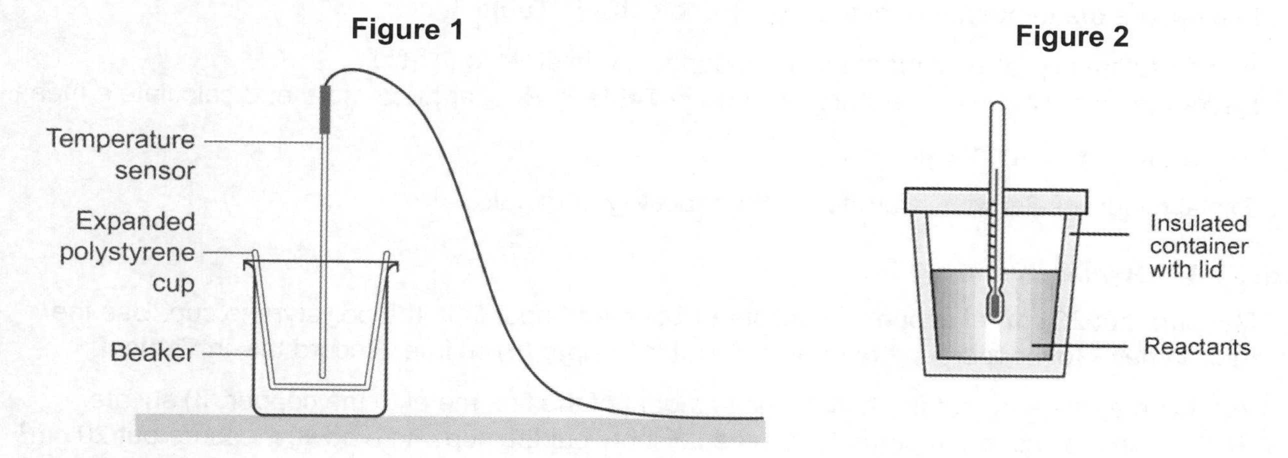

Activity 1 – Neutralisation

1. Measure 50 cm^3 of hydrochloric acid using the larger measuring cylinder and pour into the polystyrene cup. Use a thermometer to measure the temperature of the acid. Record this in **Table 1**.
2. Measure out 5 cm^3 of sodium hydroxide using the 10 cm^3 measuring cylinder and carefully add the sodium hydroxide to the acid.

   Record the maximum temperature reached in **Table 1**.
3. Repeat step **2** adding another 5 cm^3 until 40 cm^3 of sodium hydroxide has been added.

   The last few additions should produce a temperature fall rather than a temperature rise.
4. Empty the cup, measure out 50 cm^3 of fresh hydrochloric acid and repeat to gather a second set of data. Calculate the mean maximum temperature reached for each of the sodium hydroxide volumes.
5. On graph paper, plot a graph with:
   - **Mean maximum temperature in °C** on the *y*-axis
   - **Total volume of sodium hydroxide added in cm^3** on the *x*-axis.

   Draw two straight lines of best fit:
   - one through the points that are increasing
   - one through the points that are decreasing.

   Ensure the two lines are extended so they cross each other. Stick your graph in this book.
6. Use the graph to estimate how much sodium hydroxide solution was needed to neutralise 25 cm^3 of dilute hydrochloric acid.

Activity 2 – Acids + carbonates

1. Measure out 20 cm^3 of ethanoic acid and pour into the polystyrene cup. Use the thermometer to measure the temperature of the acid. Record this in **Table 2**.
2. Add one spatula of a metal carbonate to the ethanoic acid.

   Record the maximum temperature reached in **Table 2**.
3. Repeat step **2** until five spatulas of metal carbonate have been added.
4. Repeat to gather a second sat of data for a different carbonate. Compare the results for the two carbonates.

Activity 3 – Acids + metals

1. Measure out 20 cm^3 of hydrochloric acid and pour in to the polystyrene cup. Use the thermometer to measure the temperature of the acid. Record this in **Table 3**.
2. Add three small pieces of magnesium ribbon to the hydrochloric acid.
   Record the maximum temperature reached in **Table 3**. Repeat twice more and calculate a mean.
3. Repeat the experiment with zinc.
4. Explain your results with reference to the reactivity of metals.

Activity 4 – Displacement

1. Measure out 20 cm^3 of copper (II) sulfate solution and pour in to the polystyrene cup. Use the thermometer to measure the temperature of the copper (II) sulfate. Record this in **Table 4**.
2. You have a range of metals. Add a small amount of the first metal to the copper (II) sulfate. Record the maximum temperature reached. Empty out the cup into a bowl, measure out 20 cm^3 of fresh copper (II) sulfate. Repeat twice more and calculate a mean.
3. Repeat step **2** with the same amounts of other metal samples.
4. Explain your results with reference to the reactivity of metals.

Record your results

Table 1 – Neutralisation

Volume of sodium hydroxide added (cm^3)	Maximum temperature reached (°C)	
	1	2
0		
5		
10		
15		
20		
25		
30		
35		
40		

Table 2 – Acids + carbonates

Number of spatulas of carbonate added	Maximum temperature reached (°C)	
	Carbonate 1	Carbonate 2
0		
1		
2		
3		
4		
5		

Table 3 – Acids + metals

Metals	Maximum temperature reached (°C)			
	1	2	3	mean
Magnesium				
Zinc				

Table 4 – Displacement

Metals	Maximum temperature reached (°C)			
	1	2	3	mean

Check your understanding

1. A polystyrene cup is used to minimise thermal energy loss.

   a. Explain why it is important to take steps to minimise thermal energy loss. [1 mark]

   ....

   b. How else could thermal energy loss be minimised in this experiment? [1 mark]

   ....

2. The student wants to measure 5.0 cm^3 of sodium hydroxide

   a. What is the resolution of the measuring cylinder that the student has to use to accurately measure this amount? [1 mark]

   ....

   b. Explain why a repeat reading is made during this experiment. [1 mark]

   ....

Exam-style questions

1. **Table 5** shows the results of a reaction between magnesium and hydrochloric acid.

Table 5

Concentration of HCl (M)	Time taken for magnesium to disappear (s)			
	1	**2**	**3**	**mean**
0.50	140	140	141	141
0.75	101	111	102	
1.00	72	71	73	72

   a. Calculate the mean for the reaction using 0.75 M.
   Take in to account any anomalous results. [2 marks]

   ....

   ....

   b. The results for this 0.50 M and 1.00 M in this experiment are **precise**.

   Explain how the results for these two concentrations show that the experiment is precise. [1 mark]

   ....

   c. Describe the relationship between concentration of acid and time taken for the magnesium to disappear. [1 mark]

   ....

   d. Explain this relationship in terms of collision theory. [2 marks]

   ....

   ....

2. **HT** Reactions of metals with acids are exothermic. Explain why this reaction is exothermic, using the idea of chemical bonds being made and broken. [2 marks]

   ....

   ....

4.6.1.2 Rates of reaction

Collision theory helps us to understand and predict the rate of reaction when certain conditions are changed. To react, particles need to collide; they need to be in contact with each other, and they also need to collide with enough energy to react – they need to be going fast enough. We can increase the probability that they will collide by increasing the concentration.

You are going to investigate how changes in concentration affect the rates of reactions. You will measure this by measuring the volume of a gas produced in one experiment and by observing a change in turbidity (cloudiness) in another experiment.

Before you start these experiments, make a hypothesis: how will concentration affect the rate of reaction? Be sure to include an explanation based on collision theory to back up your prediction. Write your hypothesis in **Question 1a**.

Learning outcomes	Maths skills required
• Devise a hypothesis. • Measure volumes precisely to ensure your results are valid.	• Plot two variables from experimental data.

Apparatus list

Activity 1 – Measuring a change in turbidity

- 40 g/dm^3 sodium thiosulfate solution
- 2.0 M hydrochloric acid
- 10 cm^3 measuring cylinder
- 50 cm^3 measuring cylinder
- 100 cm^3 conical flask
- eye protection
- black cross on a sheet of paper (preferably printed)
- stopclock

Activity 2 – Measuring a gas produced

- eye protection
- side arm flask and gas syringe

or

- conical flask (100 cm^3), with a single-holed rubber bung and delivery tube to fit the conical flask, trough or plastic washing-up bowl and upside down 100 cm^3 measuring cylinder filled with water
- 50 cm^3 measuring cylinder
- clamp stand, boss and clamp
- stopclock
- magnesium ribbon cut into 3 cm lengths
- dilute hydrochloric acid (2.0 M and 1.0 M).

Safety notes

- Wear eye protection at all times.
- Sulfur dioxide is released during the reaction. Make sure the lab is well ventilated.

Always be careful when handling chemicals and follow your teacher's safety advice about the below:

- 2.0 M hydrochloric acid *(irritant)*
- 1.0 M hydrochloric acid *(low hazard)*
- 40 g/dm^3 sodium thiosulfate solution *(low hazard)*

Common mistakes

- If using the upside-down measuring cylinder filled with water, make sure there are no bubbles of air at the top of your measuring cylinder as this will make your results inaccurate.
- If using a gas syringe, make sure you are careful. Glass measuring syringes have very low friction between the inner syringe chamber and the surrounding glass. It will slide out **very** easily!
- Make sure you start the stopclock as soon as the reactants are added to each other.

Methods

Read these instructions carefully before you start work.

There are two activities to complete.

Activity 1 – Measuring turbidity

1. Measure 10 cm^3 sodium thiosulfate solution with a 10 cm^3 measuring cylinder and pour into a conical flask.

   Measure 40 cm^3 water with a 50 cm^3 measuring cylinder and add it to the 10 cm^3 of sodium thiosulfate in the conical flask. The diluted sodium thiosulfate will be at a concentration of 8 g/dm^3.

   Place the conical flask on the black cross and wash the 10 cm^3 measuring cylinder.
2. Measure 10 cm^3 of dilute hydrochloric acid with the clean 10 cm^3 measuring cylinder.
3. Start the stopclock **as soon as you add the acid to the conical flask**. Keep swirling the flask gently.
4. Observe the reaction by looking down through the top of the flask as shown in **Figure 1**.

   Stop the clock when you can no longer see the cross.

   Take care not to breathe in any sulfur dioxide fumes.
5. Record the time it takes for the cross to disappear in the **First trial** column of **Table 1**.

   Rinse all the equipment and do two more repeats for this concentration.

   Calculate the mean time in seconds.
6. Repeat steps **1–5** four times for the following concentrations:
   - 16 g/dm^3 (20 cm^3 sodium thiosulfate + 30 cm^3 water)
   - 24 g/dm^3 (30 cm^3 sodium thiosulfate + 20 cm^3 water)
   - 32 g/dm^3 (40 cm^3 sodium thiosulfate + 10 cm^3 water)
   - 40 g/dm^3 (50 cm^3 sodium thiosulfate + no water)
7. Plot your data on **Graph 1** and draw a smooth curved line of best fit.

Figure 1

Activity 2 – Measuring the volume of a gas produced

1. Set up the apparatus as shown in **Figure 2** <u>or</u> use a conical flask (100 cm^3) and single-holed rubber bung and delivery tube to fit the conical flask, with an upside down 100 cm^3 measuring cylinder filled with water in a trough or plastic washing up bowl.

Figure 2

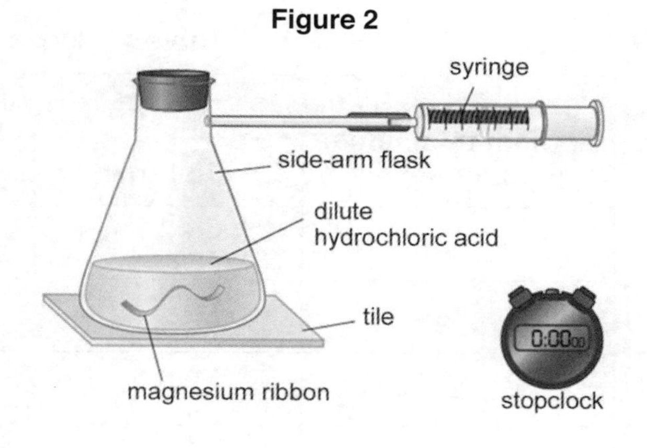

2. Measure 50 cm^3 of 2.0 M hydrochloric acid using a 50 cm^3 measuring cylinder and pour the acid into the 100 cm^3 conical flask.
3. Start the stopclock **as soon as you add the 3 cm strip of magnesium ribbon to the flask**, putting the bung back into the flask as quickly as you can.
4. Every 10 seconds, record the volume of hydrogen gas given off in **Table 2**. Continue timing until no more gas appears to be given off.
5. Repeat steps **1–4** using 1.0 M hydrochloric acid.
6. Plot your data for 2.0 M hydrochloric acid on **Graph 2** with:
   - **Volume of gas (cm^3)** on the *y*-axis
   - **Time (s)** on the *x*-axis

   Draw a smooth curved line of best fit using a solid line.
7. Plot the points for 1.0 M hydrochloric acid on the same graph and draw a smooth curved line of best fit using a dashed line or a different colour.
8. Use this graph to compare the rates of reaction of 1.0 M and 2.0 M hydrochloric acid with magnesium at 20 s. Do this by drawing a tangent to the 1.0 M curve at 20 s and compare the slope of the tangents. Repeat for the 2.0 M curve.

   HT Then calculate the gradient of the tangent to the curve.

Record your results

Table 1 – Measuring a change in turbidity

Concentration of sodium thiosulfate in g/dm^3	Time taken for cross to disappear (s)			
	First trial	Second trial	Third trial	Mean
8				
16				
24				
32				
40				

Graph 1

Table 2 – Measuring a gas produced

Time (s)	Volume of gas produced (cm^3)	
	2.0 M hydrochloric acid	1.0 M hydrochloric acid
10		
20		
30		
40		
50		
60		
70		
80		
90		
100		

Graph 2

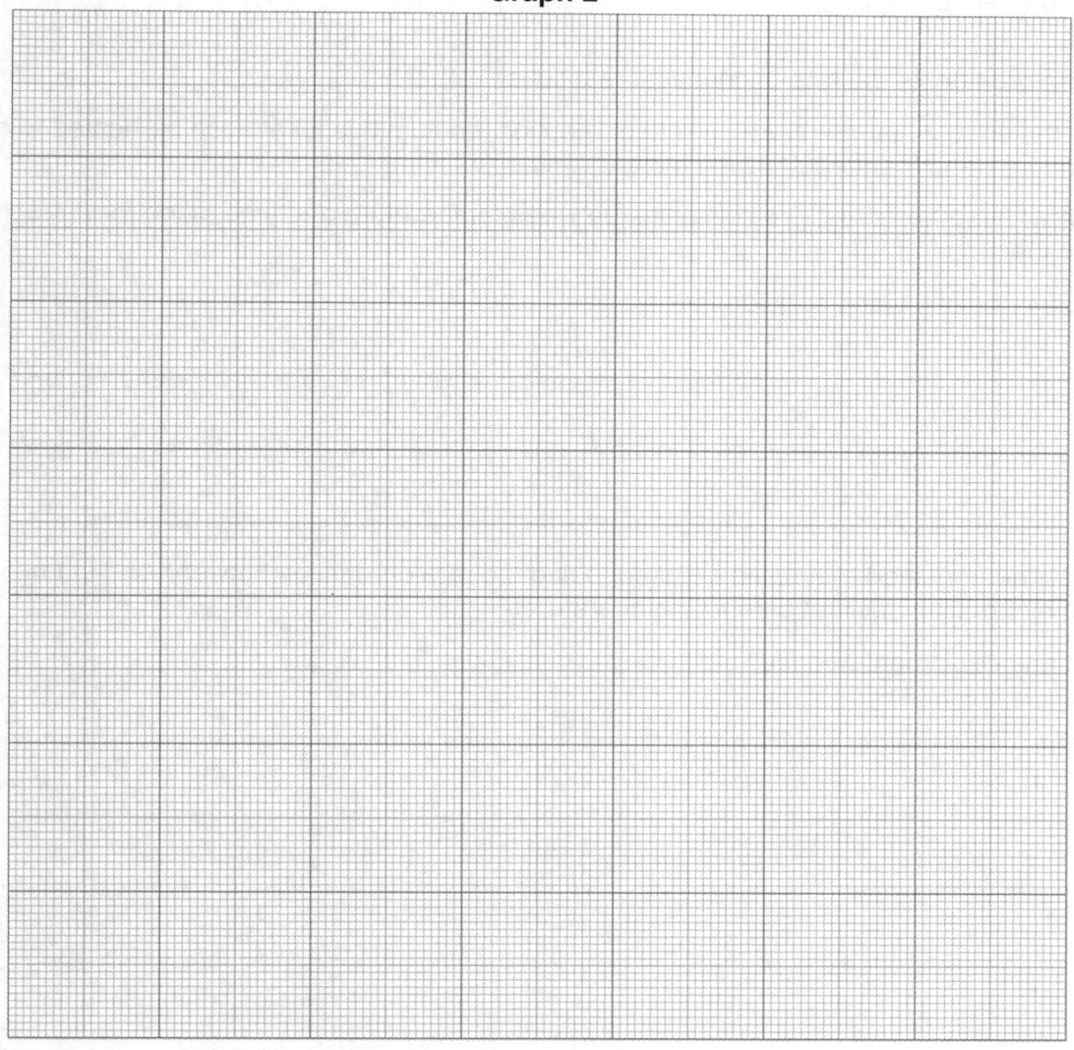

Check your understanding

1. You are changing concentration in **Activity 1** and **Activity 2**.

   a. Write a **hypothesis** that predicts a link between concentration and time taken for a reaction to take place.

   Explain the link using collision theory. [2 marks]

   ....

   ....

   b. State another factor, other than concentration, that could be changed to affect rate of reaction. [1 mark]

   ....

2. In **Activity 2** you gathered one set of data for two different concentrations of hydrochloric acid.

   Describe how you could gather more data to support your hypothesis in **Question 1**. [1 mark]

   ....

Exam-style questions

1. Temperature affects the rate of chemical reactions.

   To see the effect of temperature on rate of reaction, sodium thiosulfate can be heated to different temperatures and then reacted with hydrochloric acid.

   a. Write a hypothesis that predicts a link between temperature and time taken for a reaction to take place.

   Explain the link using collision theory. [3 marks]

   ....

   ....

   ....

   b. Describe a method that could be used to find the rate of reaction between sodium thiosulfate and hydrochloric acid at different temperatures.

   List any equipment needed. [6 marks]

   ....

   ....

   ....

   ....

   ....

....

....

....

....

2. Hydrogen peroxide (H_2O_2) decomposes to form water and oxygen.

a. Complete the equation for this reaction.

Balance the equation. [2 marks]

.... H_2O_2 → .... H_2O + ...

b. The decomposition of hydrogen peroxide can be increased by using a catalyst.

Explain why catalysts increase rates of reaction. [2 marks]

....

....

c. Describe a method that could be used to measure how much a catalyst increases the rate of reaction of hydrogen peroxide decomposition. [2 marks]

....

....

4.8.1.3 Chromatography

Paper chromatography is a useful instrumental method to allow analysis of mixtures. You will investigate how paper chromatography can be used to separate and tell the difference between coloured substances.

You will calculate R_f (retention factor) values for the substances separated. The R_f value can never be higher than 1. R_f values that are close to 1 (e.g. 0.9) show that the food colouring pigments have spent more time in the mobile phase than in the stationary phase. Lower R_f values show the food colouring pigments have spent more time in the stationary phase than in the mobile phase. (If you are able to record this experiment using time lapse, it's very satisfying to watch when sped up).

Learning outcomes	Maths skills required	Formulae
• Make and record measurements used in paper chromatography. • Calculate R_f values.	• Calculate R_f values.	• $R_f = \frac{\text{distance moved by substance}}{\text{distance moved by solvent}}$

Apparatus list

- 250 cm^3 beaker with lid
- a rectangle of chromatography paper
- deionised water (solvent)
- glass capillary tubes
- paper clips
- four known food colourings labelled **A** to **D**
- an **unknown mixture** of food colourings labelled **U**
- hairdryer

Safety notes

- Try not to snap the capillary tube. They are made of thin glass.

Common mistakes

- Make sure you draw the line at the bottom of the chromatography paper in pencil. If it is drawn with ink, it will run and ruin the chromatogram.
- Do not let the food colouring spots touch the water; otherwise the food colouring will run into the water rather than up the chromatography paper.
- Don't let the chromatography paper touch the side of the beaker. If you do, it will distort the solvent front and make it harder to calculate an accurate R_f value.
- Don't forget about it! If the food colourings run off the edge of the chromatography paper, then you cannot calculate the R_f value, as you cannot measure the solvent front.
- Also, don't take the paper out too early. If the food colouring is moving with the solvent front (in the mobile phase), let it run a little further until it stops and is left behind on the paper (the stationary phase). If you take the paper out too early, all your food colourings will travel the same distance as the solvent front, which will make all your R_f values 1.

Methods

Read these instructions carefully before you start work.

1. Draw a pencil line 1.5 cm from the bottom edge of the chromatography paper.
2. Place a spot of the unknown substance on the pencil line using the capillary tube. Place spots of other food colourings alongside using different capillary tubes. The spots need to be small and concentrated. You can add more food colouring to each spot when it is dry. Record the order of the colours on the paper.
3. Add deionised water to the beaker so that it is 1 cm deep. Place the chromatography paper in the beaker so that the water can rise up the chromatogram (see **Figure 1**).

   The spots of food colouring **must** be above the water level. You can use the paper clips to hold the paper in place.

Figure 1

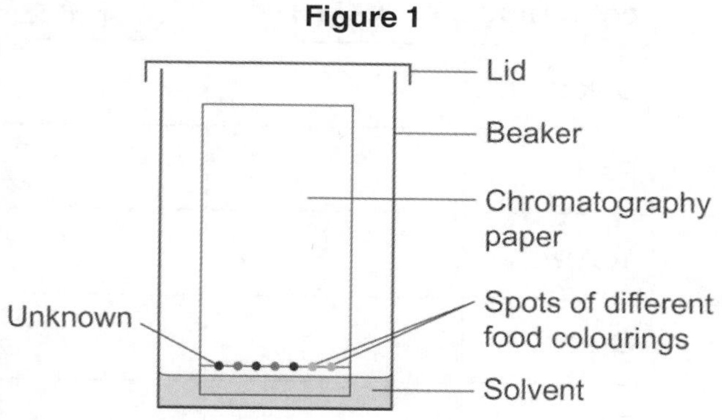

4. Allow the solvent (water) to rise up the chromatography paper until it is almost at the top. Remove the paper from the solvent and use a hairdryer to dry the chromatogram. Alternatively, you can leave them on a radiator.
5. Measure:
   - the distance from the pencil line to the solvent front
   - the distance from the pencil line to the middle of each spot.

   Your food colouring may separate out into a number of spots – you have space to record five spots. There may only be one or two spots. In this case, leave the other boxes blank.

   Record these values in **Table 1**.

Record your results

1. Distance/cm moved by the solvent = ...

Table 1 – Distance moved by food colourings

Food colouring	Distance moved by spot (cm)				
	Spot 1	Spot 2	Spot 3	Spot 4	Spot 5
Unknown					
Red					
Yellow					
Green					
Blue					

2. Calculate the R_f value for each spot on the chromatogram.

$$R_f \text{ value} = \frac{\text{distance moved by substance}}{\text{distance moved by solvent}}$$

(Each value should be between 0 and 1.)

Record these values in **Table 2**.

Table 2 – Calculating R_f values

Food colouring	R_f value of food colouring spots				
	Spot 1	Spot 2	Spot 3	Spot 4	Spot 5
Unknown					
Red					
Yellow					
Green					
Blue					

3. When dry, attach your chromatogram in the space below.

Check your understanding

1. A substance has an R_f value of 0.9.

   a. On the chromatogram, would you expect to see this substance at the top near the solvent front **or** at the bottom near the pencil line? [1 mark]

   ....

b. Describe the solubility of this substance. [1 mark]

....

2. Another substance has an R_f value of 0.

a. On the chromatogram, would you expect to see this substance at the top near the solvent front **or** at the bottom near the pencil line? [1 mark]

....

b. Describe the solubility of this substance. [1 mark]

....

3. Why is it important to add a lid to your chromatography beaker? [1 mark]

....

4. A student uses five different permanent markers for a chromatogram. She uses water as the solvent.

After running the chromatograph, the student calculates that **all** the permanent markers have an R_f value of 0.

Suggest two methods to improve her experiment. [2 marks]

....

....

Exam-style questions

1. Soy sauce is a mixture of different amino acids.

Thin layer chromatography is a technique that can be used to separate and identify amino acids.

The R_f values of different amino acids are listed in **Table 3**.

Table 3

Amino acid	R_f value
Alanine	0.30
Cystine	0.14
Phenylalanine	0.62
Serine	0.26

a. Which amino acid spends the most time in the mobile phase? [1 mark]

....

b. Which amino acid is most soluble in the solvent used in this experiment? [1 mark]

....

2. R_f values fall between 0 and 1.

a. Explain why an R_f value cannot be greater than 1. [1 mark]

....

Figure 2 shows a chromatogram of E numbers in food.

The R_f value of E131 is 0.9.

Figure 2

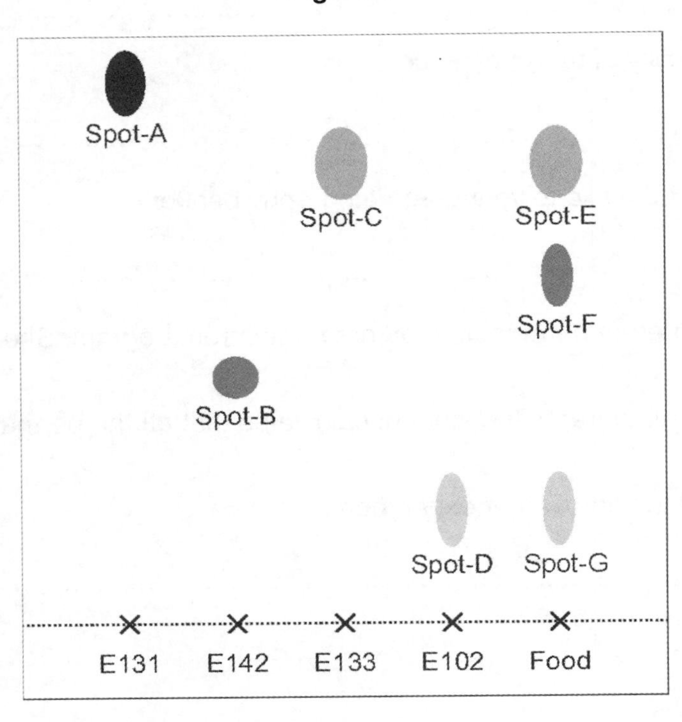

b. Calculate the R_f values for each of the spots in the food sample. [3 marks]

....

....

....

c. Which E numbers can you positively identify as being in the food sample? [2 marks]

....

....

d. Suggest how the experiment could be improved to identify the unknown sample of food. [3 marks]

....

....

....

4.8.3.7 Identifying ions

Ion tests are useful in identifying unknown substances. Flame tests, addition of sodium hydroxide, addition of acids, addition of barium chloride and addition of silver nitrate all give clues about the compound being tested. Bringing these clues together allows us to deduce what ions an unknown ionic compound contains.

In this practical you will analyse a range of known ionic compounds to find positive test results for different salts and apply this knowledge to identify an unknown compound. This practical may take more than one lesson.

Learning outcomes	Maths skills required
• Understand how to carry out flame tests and precipitate hydroxides to identify cations and how to carry out test tube reactions to identify carbonate, sulfate and halide anions. • Carry out the above identification tests safely and effectively. • Use the results of identification tests to identify the ions present in a single ionic compound.	• Estimate volumes.

Apparatus list

- Bunsen burner
- test tubes and test tube rack
- teat pipette
- nichrome wire mounted in handle
- limewater
- concentrated hydrochloric acid for cleaning nichrome wire
- 0.4 M dilute hydrochloric acid
- 0.4 M dilute sodium hydroxide
- 0.1 M barium chloride solution
- 0.4 M dilute nitric acid
- 0.05 M silver nitrate solution
- known labelled solutions: chlorides of lithium, sodium, potassium, calcium and copper
- known labelled solutions: sodium salts containing carbonate, sulfate, chloride, bromide and iodide
- salt solution labelled 'unknown'

Safety notes

- Wear eye protection at all times.

Always be careful when handling chemicals and follow your teacher's safety advice about the below:

- 0.4 M dilute hydrochloric acid *(low hazard)*
- 0.4 M dilute sodium hydroxide *(irritant)*
- 0.1 M barium chloride solution *(harmful)*
- 0.4 M dilute nitric acid *(irritant)*
- 0.05 M silver nitrate solution *(low hazard)*
- lithium chloride *(low hazard unless saturated)*
- potassium chloride *(low hazard)*
- calcium chloride *(low hazard unless saturated)*
- sodium chloride *(low hazard)*
- copper chloride *(1.4 M or more irritant, less than 1.4 M low hazard)*
- sodium carbonate *(low hazard unless saturated)*
- sodium sulfate *(low hazard)*
- sodium bromide *(low hazard)*
- sodium iodide *(low hazard)*

Common mistakes

- Don't forget to clean your nichrome wire; otherwise you will have contamination and inconclusive results.
- Make sure you record your results as you work as there are a lot of ions to test.
- Make sure you don't mix up the substances, if you are unsure, ask your teacher. You can label these using stickers or record the positions of the substances in your test tube rack.

Methods

Read these instructions carefully before you start work.

There are six activities to complete.

Activity 1 – Flame tests

1. Pour around 1 cm depth of
   - **a.** lithium chloride
   - **b.** sodium chloride
   - **c.** potassium chloride
   - **d.** calcium chloride
   - **e.** copper chloride

   into five test tubes in a test tube rack.
2. Dip the nichrome wire into the first solution. Then hold the tip of the wire in a blue Bunsen burner flame.
3. Record your observation in the **Flame test** row of **Table 1**.
4. Clean the wire carefully by dipping it in to concentrated hydrochloric acid and then holding it in the flame until no colour appears.
5. Repeat steps **2–4** for each of the other four solutions.
6. Empty and clean the test tubes.

Activity 2 – Hydroxide tests

1. Pour around 1 cm depth of
   - **a.** lithium chloride
   - **b.** sodium chloride
   - **c.** potassium chloride
   - **d.** calcium chloride
   - **e.** copper chloride

   into five test tubes in a test tube rack.
2. Add a little **sodium hydroxide** and record any precipitation.
3. Record your observation in the **Hydroxide test** row of **Table 1**.
4. Add sodium hydroxide in excess and observe if the precipitate dissolves completely.
5. Repeat steps **2–4** for each of the other four solutions.
6. Empty and clean the test tubes.

Activity 3 – Carbonate test

1. Pour around 1 cm depth of
   - **a.** sodium carbonate
   - **b.** sodium sulfate
   - **c.** sodium chloride
   - **d.** sodium bromide
   - **e.** sodium iodide

   into five test tubes in a test tube rack.
2. Add approximately 1 cm depth of **dilute hydrochloric acid** to each sodium salt in turn.
3. **If you see bubbles** – use the teat pipette to transfer the gas produced to a test tube of limewater.
4. You will need to take several pipettes of the gas to get a change in the limewater.
5. Record your results in the **Carbonate test** row of **Table 2.**
6. Empty and clean the test tubes.

Activity 4 – Sulfate test

1. Pour around 1 cm depth of
   - **a.** sodium carbonate
   - **b.** sodium sulfate
   - **c.** sodium chloride
   - **d.** sodium bromide
   - **e.** sodium iodide

   into five test tubes in a test tube rack.
2. Add a few drops of **dilute hydrochloric acid** to each solution.
   Add 1 cm depth of **barium chloride** solution.
3. Record your observations in the **Sulfate test** row of **Table 2**.
4. Empty and clean the test tubes.

Activity 5 – Halide test

1. Pour around 1 cm depth of
   - **a.** sodium carbonate
   - **b.** sodium sulfate
   - **c.** sodium chloride
   - **d.** sodium bromide
   - **e.** sodium iodide

   into five test tubes in a test tube rack.
2. Add a few drops of **dilute nitric acid** to each solution.
   Add 1 cm depth of **silver nitrate** solution.
3. Record your observations in the **Halide test** row of **Table 2**.

Activity 6 – Unknown

1. Repeat each of the flame, hydroxide, carbonate, sulfate and halide tests on the unknown salt solution.
2. Identify the unknown salt.
3. Record the resutls in **Table 3**.

Record your results

Table 1 – Metal ions (cation)

- Possible flame colours: green, crimson, lilac, yellow, orange-red.
- Possible hydroxide precipitate colours: white precipitate (which may or may not dissolve in excess sodium hydroxide), blue precipitate.

Metal ion	Lithium	Sodium	Potassium	Calcium	Copper
Flame test					
Hydroxide test					

Table 2 – Non-metal ions (anion)

- Possible products: carbon dioxide bubbles, white, cream or yellow precipitates

Non-metal ion	Carbonate	Sulfate	Chloride	Bromide	Iodide
Carbonate test					
Sulfate test					
Halide test					

Table 3 – Unknown salt

Test	Observation	Identity
Flame test		
Hydroxide test		
Carbonate test		
Sulfate test		
Halide test		

Check your understanding

1. A student is performing a flame test to identify a metal cation.

   The flame test shows a green and yellow flame.

   a. Suggest a reason for this observation. [1 mark]

   ....

   b. Describe what the student should do to gain an accurate result. [2 marks]

   ....

   ....

2. For each salt, state the ion tests **and** the result you would observe.

   a. Copper sulfate [3 marks]

      i. two tests for copper ions ....

      ....

      ....

      ii. one test for sulfate ions ....

      ....

   b. Calcium carbonate [3 marks]

      i. two tests for calcium ions ....

      ....

      ....

      ii. one test for carbonate ions ....

      ....

   c. Potassium iodide [2 marks]

      i. one test for potassium ions ....

      ....

      ii. one test for iodine ions ....

      ....

Exam-style questions

1. A student tests an unknown salt using a range of ion tests.

   The results of these tests are shown in **Table 4**.

Table 4

Test	Observation
Flame test	Crimson flame
Hydroxide test	No colour change
Carbonate test	No gas produced
Sulfate test	No colour change
Halide test	Yellow precipitate

   a. Identify the unknown salt. [2 marks]

   ....

   ....

   b. Explain your reasoning. [2 marks]

   ....

   ....

2. A student suspects that an unknown substance is sodium carbonate.
   Describe a series of tests the student could perform to positively identify the substance as sodium carbonate. [4 marks]

   ....

   ....

   ....

   ....

3. A student is performing ion tests to identify an unknown salt. When dilute hydrochloric acid is added, a gas is produced.

   a. Describe how the student could confirm that the gas produced is carbon dioxide. [1 mark]

   ....

   b. Suggest what the non-metal anion in the unknown salt could be. [1 mark]

   ....

4.10.1.2 Water purification

Water that is safe to drink is called potable water. In a chemical sense, potable water is not pure water because it can contain dissolved substances that are needed by the body. Some countries do not have access to fresh water, only sea water. This has dissolved salts in it that need to be removed. This is done by distillation.

In this investigation you will test spring water, sea water and rain water for pH and the presence of dissolved salts. After distillation of the sea water, you will test the water again to check that dissolved salts have been removed. If they have, the water is fit for drinking! Don't actually drink it though! Nothing prepared in a school science lab is safe for human consumption. This practical may take more than one lesson.

Learning outcomes	Maths skills required
• Safely purify water. • Compare pH and dissolved salts of a range of water samples. • Correctly use the apparatus, including the water bath.	• Estimate of volumes • Calculate mass

Apparatus list

- 50 cm^3 sample of 'sea water'*
- 10 cm^3 sample of 'spring water'*
- 10 cm^3 sample of 'rain water'*
- 10 cm^3 sample of 'sea water after distillation'
- universal indicator
- test tubes and rack
- Bunsen burner
- eye protection
- 10 cm^3 measuring cylinder
- 50 cm^3 measuring cylinder
- tripod
- gauze
- heatproof mat
- 250 cm^3 beaker
- watch glass
- tongs
- clamp stand
- 250 cm^3 conical flask
- delivery tube with bung
- anti-bumping granules
- ice
- balance

*Technician notes for creating these samples can be found on the Collins website.

Safety notes

Activity 2

- Wear eye protection.
- **Do not** let the water bath boil dry!

Activity 3

- Wear eye protection.
- Make sure you have enough water in the conical flask to stop it boiling dry and cracking.
- Don't remove the conical flask from the heat during distillation! If the sea water starts to boil over, reduce the heat but do not remove it. This is to prevent suck back.
- Make sure you keep the delivery tube **at least 2 cm away** from the distilled water in the test tube. If you stop heating, the cold distilled water will be sucked back into the hot conical flask and can cause it to crack.

Common mistakes

- If you do not produce enough salts to measure a mass, you can make a qualitative observation of how much salt there is and the appearance of the salt. Just don't write down a mass of 0 g where there is obviously a residue there!
- When adding drops of universal indicator, add one or two drops. Too much universal indicator makes the colour hard to identify.

Methods

Read these instructions carefully before you start work.

There are three activities to complete.

(Your teacher might demonstrate the distillation of seawater to obtain water using a Liebig condenser rather than you doing **Activity 3**. You can use this distilled sea water as one of your samples rather than obtaining your own.)

Activity 1 – pH

1. Pour around 1 cm depth of
   a. sea water
   b. spring water
   c. rain water
   d. sea water after distillation
   into a test tube in the rack.
2. Add a one or two drops of universal indicator solution to each of the samples. Using a pH colour chart, match the colour and record the pH of the water in the **pH** column of **Table 1**.

Activity 2 – Collecting dissolved salts

1. Weigh a clean, dry watch glass. Record its mass in **Table 1**.
2. Set up your Bunsen burner on a heatproof mat with a tripod and gauze over it. Place the beaker with the water on the gauze (see **Figure 1**).
3. Measure 4 cm^3 of one of your water samples using a 10 cm^3 measuring cylinder and place it in a watch glass above a beaker with approxomately 200 cm^3 of tap water in it (see **Figure 1**). This beaker will act as a water bath to make sure your water sample will not boil too quickly.
4. Heat the sample until all the water has evaporated from the watch glass and the dissolved salts are left behind. **Do not let the water bath boil dry**.
5. Remove the watch glass with tongs and allow to cool. Dry the underside of the watch glass with a paper towel to remove any water.

   Measure the mass of the watch glass with the salts and record the mass in **Table 1**. Calculate the mass of the dissolved solids. Wash the watch glass and dry it.

Figure 1

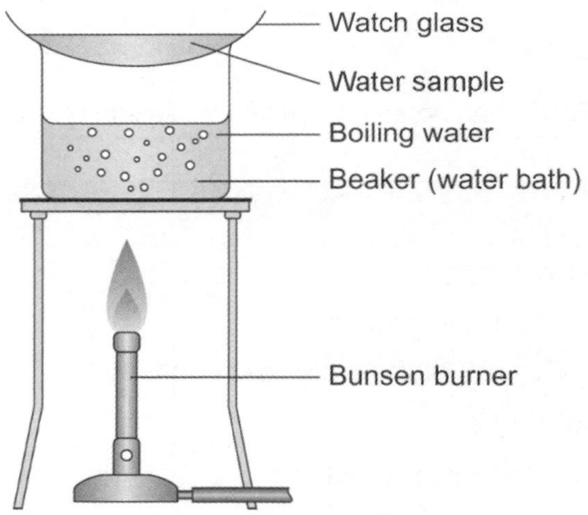

6. Repeat steps **3–5** for the other water samples. (If you use the same watch glass each time and you have cleaned it well, you do not need to measure the mass of the empty watch glass again.)

Activity 3 – Desalination of sea water

1. Pour the remaining salt solution into the conical flask. Add a few anti-bumping granules and set up the apparatus as in **Figure 2**.
2. Clamp the neck of the conical flask securely. Half-fill the beaker with ice and water.
3. Heat the salt solution with a blue Bunsen burner flame until it starts to boil. Adjust the flame so that the salt solution boils gently.

   Water will condense in the test tube. Collect about a 2 cm depth of distilled water.
4. Repeat **Activity 1** and **Activity 2** for your sample of distilled water
5. If you have been provided with different salt solution samples, you will need to repeat the procedures on these.

Figure 2

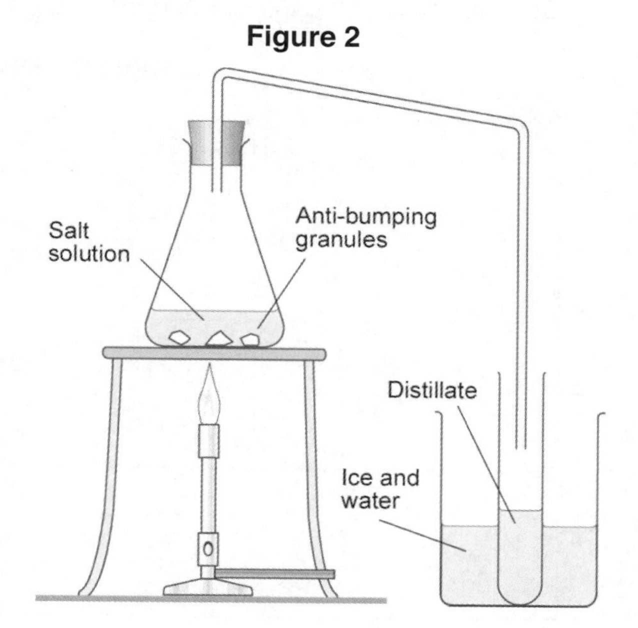

Record your results

Table 1 – pH of sample and mass of dissolved solids

Water sample	pH	Mass (g)		
		Watch glass	Watch glass with dissolved solids	Dissolved solids
Sea				
Spring				
Rain				
Sea water after distillation				

Check your understanding

1. In the table below, list three risks that need to be controlled during this experiment.

   Describe how you can control each of them. Write your answers in **Table 2**. [6 marks]

Table 2 – Risks and control methods

Risk	Control method

2. The water provided for this experiment came from only one source.

   Explain how you could make your results a more accurate representation of the salts present in different sources of water from across the whole of the UK. [3 marks]

   ....

   ....

   ....

3. Obtaining pure drinking water by the distillation of sea water is possible, but is rarely carried out.

   Suggest why distillation of sea water might be hard to achieve on a large enough scale to provide drinking water for a community. [2 marks]

   ....

   ....

Exam-style questions

1. A sample of 10 cm^3 of sea water is heated and evaporated in a watch glass to obtain the dissolved salts.

   The results are shown in **Table 3**.

Table 3

Water sample	Mass (g)		
	Watch glass	Watch glass with dissolved solids	Dissolved solids
Sea	10.97	11.06	

   a. Calculate the mass of the dissolved solids. [1 mark]

   ....

   b. Suggests how the accuracy of this experiment could be improved. [2 marks]

   ....

   ....

2. Desalination is the process of removing excess salts from salt water or sea water to provide potable water.

   a. Describe a method for producing potable water from sea water in a laboratory. [6 marks]

   ....

   ....

   ....

   ....

   ....

   ....

   b. Explain why potable water is **not** pure water. [1 mark]

   ....

   ....

Answers

For 6 mark method questions you will need to consider:
0 marks: No relevant content | **1-2 marks:** Simple statements are made, Some understanding is demonstrated, Some scientific techniques and procedures are relevant, Lacks logical structure, Valid results cannot be produced | **3-4 marks:** Majority of method is present in detail, Reasonable understanding is demonstrated, Most scientific techniques and procedures are relevant, Mostly logical sequence but some may be illogical and not detailed, Valid results may be produced | **5-6 marks:** Coherent method is present in detail, Good understanding is demonstrated, Broad understanding of scientific techniques and procedures, Logical sequence to method, Valid results can be produced

4.4.2.3 Making salts

Check your understanding

1. **a.** There is a greater amount of a reactant than necessary to react completely with the limiting reactant. [1]
   b. Harmful sulfur dioxide gas could be produced when the mixture is heated. [1]
2. **a.** A_r Cu = 63.5
   A_r O = 16
   M_r CuO = 63.5 + 16 = 79.5 [1]
   for calculating M_r of CuO

   A_r H = 1
   A_r S = 32
   A_r O = 16
   M_r H_2SO_4 = (1 x 2) + 32 + (16 x 4) = 98 [1]
   for calculating M_r of H_2SO_4

   M_r reactants = 79.5 + 98 = 177.5 [1]
   for adding M_r of CuO and M_r of H_2SO_4.
   (Allow mark for error from previous M_r calculations carried forward.)
   b. No atoms are lost or made during a chemical reaction (so the mass of the products equals the mass of the reactants). [1]
   c. The mass of the products equals the mass of the reactants. [1]
   The total mass of the products should be 177.5, or carry forward student's answer from (**a**). [1]

Exam-style questions

1. **a.** The crystals are copper chloride. [1]
   b. Answer should include: [6]
   - heat hydrochloric acid
   - add copper oxide and stir until it is in excess
   - filter to remove unreacted copper oxide
   - place the solution in an evaporating basin over a beaker acting as a water bath
   - evaporate at least half the liquid
   - check for the crystallisation point using a cool glass rod
   - pour the solution into a petri dish
   - allow it to crystallise.
2. **a.** $CaCO_3 + 2HCl \longrightarrow CaCl_2 + H_2O + CO_2$ [1]
   b. calcium chloride [1]
3. **a.** sodium and nitric acid [1]
   b. sodium is very reactive/it could explode [1]

4.4.2.4 Neutralisation

Check your understanding

1. **a.** how close to the true value a result is [1]
   b. where there is very little spread around the mean / if they cluster together closely [1]
   c. a correct comment on the precision of the results [1]
   d. Burettes allow precise measurement of liquids as they have intervals of 0.1 cm^3, OR burettes allow very small amounts to be added (dropwise). [1 for either]

Exam-style questions

1. **a.** 41.8 – 15.2 = 26.6 (cm^3) [1]
   b. (26.6 + 26.7 + 26.6)/3 = 26.6 (cm^3) [1]
   c. (26.7 – 26.6)/2 = ±0.05 (cm^3) [1]
2. **HT** Moles of NaOH:
   25.0 ÷ 1000 = 0.025
   0.5 = mol ÷ 0.025
   mol = 0.5 x 0.025
   mol = 0.0125 [1]

   Balance equation:
   $H_2SO_4 + 2NaOH \rightarrow Na_2SO_4 + 2H_2O$
   1:2 ratio
   0.0125 mol NaOH to 0.00625 mol H_2SO_4 [1]

   Calculate concentration of acid:
   Concentration (mol/dm^3) =
   number of moles ÷ volume of solution (dm^3)
   26.6 ÷ 1000 = 0.0266
   Conc = 0.00625 ÷ 0.0266
   Conc = 0.23 M [1]

4.4.3.4 Electrolysis

Check your understanding

1. copper at the cathode/negative electrode
   chlorine gas at the anode/positive electrode [1, both required]
2. **a.** $CuCl_2$ [1]
   b. 2 [1]
   c. reduced [1]
3. **a.** Use a burning splint held at the open end of a test tube of the gas. Hydrogen burns rapidly with a pop sound. [1]
   Use damp litmus paper. When damp litmus paper is put into chlorine gas, the litmus paper is bleached and turns white. [1]
   b. Any two from:
   - wear eye protection
   - use a low potential difference (to make sure excessive chlorine is not produced)
   - do not run electrolysis for more than five minutes (to make sure excessive chlorine is not produced)
   - any other sensible precaution. [2]

Exam-style questions

1. **a.** chlorine (anode) [1]

b. Iodine would be formed at the anode (as I^- is a negative ion). [1]
Hydrogen would be formed at the cathode (as H^+ is a positive ion). [1]
Hydrogen is made rather than potassium because potassium is more reactive than hydrogen, so K^+ ion would stay in solution. [1]
Iodine is made rather than oxygen as iodine is a halogen (group 7) so stays in solution. [1]
c. potassium hydroxide [1]

2. a. At the negative electrode/anode, a Cu^{2+} ion is reduced and becomes a Cu atom. [1]
$Cu^{2+} + 2e^- \rightarrow Cu$ [1]
b. When solid, bromine and lead ions are held in a lattice and cannot move. [1]
When molten, the ions can move. [1]
c. Aqueous lead bromide is dissolved in water. [1]
There are hydrogen ions in water. [1]
Hydrogen is lower than lead in the reactivity series. [1]

4.5.1.1 Temperature changes

Check your understanding

1. a. It is important as otherwise the temperature rise will be lower than the true value OR the experiment will not be accurate. [1]
b. By adding a lid to the polystyrene cup OR stirring OR any other sensible suggestion. [1]
2. a. $0.1 cm^3$ [1]
b. Repeats can be used to discount anomalous results OR repeats can be used to calculate a mean. (It is not enough to say that repeats might make it more accurate.) [1]

Exam-style questions

1. a. $(21 + 22)/2$ [1]
$= 21.5$ (s) [1]
If you didn't spot the anomaly and calculated 23 [1]
b. Any one from: the results are all similar / close together / spread over a small range. [1]
c. Either:
   - the more concentrated the acid, the larger the maximum temperature change OR
   - the more dilute the acid, the smaller the maximum temperature change. [1]

d. The particles are closer together in an acid with a higher concentration OR there are more particles in a smaller volume. [1]
More particles are more likely to collide (so reactions are more likely to occur). [1]
2. HT The energy required to break the bonds of the reactants is less than the energy released when the bonds of the products are formed. [1]
The net energy change is a transfer of energy to the environment. [1]

4.6.1.2 Rates of reaction

Check your understanding

1. a. As concentration increases, time taken to react decreases. [1]
As there are more particles per unit volume because of the increase in concentration, the frequency of collision increases. [1]
b. One from:
   - temperature
   - addition of catalyst
   - surface area
   - pressure. [1]

2. You could repeat the experiment at different concentrations. [1]

Exam-style questions

1. a. As temperature increases, rate of reaction increases. [1]
Increasing the temperature increases the frequency of collisions and makes the collisions more energetic. [1]
There are more successful collisions / more collisions have the activation energy. [1]
b. Answer should include: [6]
   - measure out the same amount of sodium thiosulfate using a **measuring cylinder**
   - heat/cool to different temperature by adding to **ice bath** and **water bath**
   - measure the temperature using a **thermometer**
   - add an amount of hydrochloric acid to the sodium thiosulfate
   - start the **stopclock** as the acid is added
   - time until the **black cross** cannot be seen
   - repeat this experiment for different temperatures
   - plot the results on a graph
   - draw a tangent to the graph and calculate the gradient
   - rate of reaction = 1/time taken for cross to disappear.

2. a. $2 H_2O_2 \rightarrow 2 H_2O + O_2$
1 mark for O_2 [1]
1 mark for correct balancing [1]
b. Catalysts increase rate of reaction by providing a different pathway for the reaction [1]
that has a lower activation energy. [1]
c. 1 mark for a valid way to measure rate (e.g. gas production or mass lost) [1]
1 mark for repeating with catalyst added [1]

4.8.1.3 Chromatography

Check your understanding

1. a. near the solvent front [1]
b. This substance is very soluble in this solvent OR it spends more time in the mobile phase (dissolved in the solvent) than it does in the stationary phase (bound to the chromatography paper). [1]
2. a. near the pencil line [1]
b. This substance is not at all soluble OR it has spent all of the time in the stationary phase (bound to the chromatography paper) and no time in the mobile phase. [1]
3. The solvent could evaporate. [1]
4. The student could either use pens that are soluble in water [1]

or they could change the solvent for one that is appropriate for permanent markers, i.e. one that permanent markers will dissolve in, such as alcohol. [1]

Exam-style questions

1. a. phenylalanine [1]
   b. phenylalanine [1]
2. a. A substance cannot travel further than the solvent front OR the distance travelled by the spot/the distance travelled by the substance cannot be greater than 1. [1]
   b. Allow answers between the following ranges:
   spot G: R_f value of 0.10 – 0.25 [1]
   spot F: R_f value of 0.50 – 0.65 [1]
   spot E: R_f value of 0.70 – 0.85 [1]
   c. E102 [1]
   E133 [1]
   d. Test a large range of samples of known E number food additives. [1]
   Compare this to the unknown sample on the chromatogram. [1]
   Calculate the R_f values and match them. [1]
   (If no other mark awarded, give the following 1 mark:)
   Look up known R_f value for E numbers [1]

4.8.3.7 Identifying ions

Check your understanding

1. a. There could be contamination of the sample / the equipment might not be clean. [1]
   b. They should clean all the equipment [1]
   using concentrated hydrochloric acid. [1]
2. a. Copper sulfate:
   - test for metal cations – flame test – positive = green flame [1]
   - test for metal cations – precipitate with sodium hydroxide – positive = blue precipitate [1]
   - test for non-metal anion – sulfate test with drops of hydrochloric acid and then add barium chloride – positive = white precipitate [1]

   b. Calcium carbonate:
   - test for metal cations – flame test – positive = orange-red flame [1]
   - test for metal cations – precipitate with sodium hydroxide – positive = white [1]
   - test for non-metal anion – carbonate test with dilute hydrochloric acid – positive = gas produced that reacts with limewater to produce cloudy precipitate [1]

   c. Potassium iodide:
   - test for metal cations – flame test – positive = lilac flame [1]
   - test for non-metal anion – halide test with silver nitrate – positive = yellow precipitate [1]

Exam-style questions

1. a. lithium iodide [2]
   b. Metal cation is lithium as flame test is a crimson flame. [1]
   Not lithium carbonate as no bubbling;
   not lithium sulfate as no colour change of white precipitate with barium chloride;
   non metal anion is iodide as yellow precipitate with silver nitrate. [1]
2. Flame test to positively identify sodium = yellow flame. [1]
   Sodium hydroxide precipitate test to discount iron, copper, calcium, magnesium and aluminium. [1]
   Carbonate test with dilute hydrochloric acid – positively. [1]
   Collect gas and test with limewater – positive result shows carbon dioxide = anion is carbonate. [1]
3. a. test by bubbling through limewater [1]
   b. carbonate [1]

4.10.1.2 Water purification

Check your understanding

1. Any valid risk and control method can be accepted examples of expected answers are:

Risk	Control method
Using a Bunsen burner and hot equipment	Use tongs for hot equipment Ensure any loose clothing and long hair tied back
The distillate/distilled water getting sucked back in to the hot conical flask causing it to crack	Heat the conical flask constantly Make sure the delivery tube is above the level of the distillate/distilled water
Getting salt water in eyes	Wear eye protection

2. More samples should be taken [1]
   from different locations. [1]
   Repeats should be conducted (and a mean should be calculated). [1]
3. For an entire community, distilling sea water would require a vast amount of energy [1]
   and would take a long time. [1]

Exam-style questions

1. a. 11.06 – 10.97 = 0.09 (g) [1]
   b. (At least) two more repeats could be done [1]
   and a mean could be calculated. [1]
2. a. Answer should include: [6]
   - use a conical flask with a bung and delivery tube
   - pour the sea water into the conical flask and heat with a Bunsen burner
   - until it evaporates
   - condense the evaporated water
   - using a cold test tube in icy water
   - using a (Liebig) condenser
   - collect the distillate/distilled water in the test tube/beaker

   b. Potable water contains dissolved substances (but pure water only contains water molecules). [1]